Made in the USA
Las Vegas, NV
06 May 2024

89576239R00070